CONTENTS

CANDLEWICK PRESS
CAMBRIDGE, MASSACHUSETTS

WINGS, STINGS AND WRIGGLY THINGS

Martin Jenkins

Although they seem smooth and glossy, butterfly wings look bumpy and rough under a microscope. They're covered in thousands of tiny scales, which overlap like roof tiles.

7 As you might expect, this giant butterfly's caterpillars feed on a giant plant. They eat only the leaves of a huge vine called a Dutchman's pipe.

8 Although the Dutchman's pipe is poisonous, it doesn't harm the birdwing's caterpillars. They store the poison in their body — and anything that tries to eat them is in for a bad shock!

5 Male Queen Alexandra's birdwings are quite different from the females. They're smaller, and their wings are a beautiful blue-green.

Butterflies sip sweet liquid nectar from flowers, using their proboscis like a bendy straw. When a butterfly isn't feeding, its proboscis is neatly coiled under its chin.

6 Lots of butterflies have tastebuds on their feet! They can tell if a flower has nectar as soon as they land on it.

3 BUTTERFLIES

1 Every butterfly starts life as a tiny egg, which hatches into a caterpillar. A caterpillar is really good at one thing—eating. It eats and eats and eats, mostly tender young leaves and flower buds.

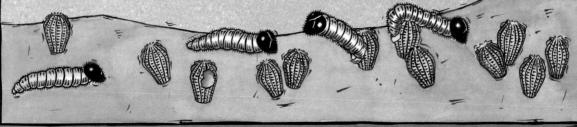

2

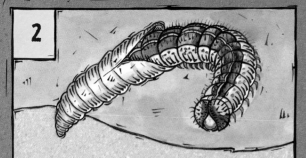

The caterpillar gets fatter and fatter, and just when you think it's close to bursting, it does! Fortunately, it's already grown a new skin. This is looser than the old skin, with plenty of room for growing.

3 The caterpillar sheds its skin four or five times, getting bigger each time. But when its skin bursts for the very last time, something totally different comes out—a chrysalis.

4 The chrysalis doesn't eat and it doesn't move, but a really weird change takes place inside it. Everything melts down into a kind of gloop— everything, that is, except for a few tiny parts. These parts slowly change and grow into legs, wings, eyes . . .

5 Until, one day, the chrysalis splits open and a damp crumpled shape crawls out. As it dries in the warm sunshine, its wings slowly unfurl. Then, all of a sudden, it takes to the air—a brand-new butterfly!

2 BUTTERFLIES

FANCY FLIERS

1 This is no ordinary insect. It's a female QUEEN ALEXANDRA'S BIRDWING, and it's the biggest butterfly in the world!

2 If this butterfly sat on your nose, its outspread wings would touch your ears — they can be almost 8 inches across from tip to tip.

3 The world's smallest butterfly is the SOUTH AFRICAN DWARF BLUE. Each wing is about the size of your little fingernail.

4 Queen Alexandra's birdwings are found in only one tiny part of Papua New Guinea. Their home is a hot, steamy rain forest.

BUTTERFLIES 5

SLIMY

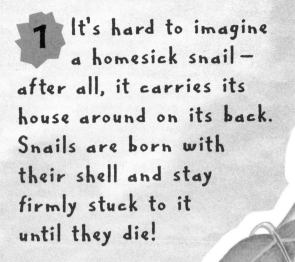

1 It's hard to imagine a homesick snail— after all, it carries its house around on its back. Snails are born with their shell and stay firmly stuck to it until they die!

2 A snail has only one foot to get around on. It slowly glides along by rippling the sole of its foot.

3 You can tell where a snail has been by its trail of silvery slime. The slime protects the snail's foot from scratches.

1 Snails are pretty unusual when it comes to making babies. They are hermaphrodites, which means that each snail is both male and female. You still need two of them to make babies, though!

2 Many kinds of snails spend hours courting before they mate. They slowly twist around each other, covering themselves in frothy slime.

3 After mating, each snail hunts for soft ground where it can dig a hole and lay its eggs. Then it seals up the hole, leaving the eggs hidden inside.

6 SNAILS

TRAILS

4 Snails have a thing called a radula in their mouth for grinding up their food. It's like a hard tongue covered with thousands of tiny teeth.

6 Lots of snails have their eyes at the end of long stalks. If a snail gets scared, it can pull its eyestalks right back inside its head.

5 Most snails are vegetarians, but they're not generally fussy eaters. Some are even fond of soggy cardboard.

These beautifully colored snails are called LITTLE AGATE SHELLS. They live on the Hawaiian island of Oahu, in the Pacific Ocean.

4 When the baby snails hatch, the first thing on their mind is food. They eat what's left of their eggshell and then look around for something else. Often they'll eat any eggs that haven't yet hatched.

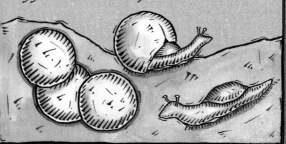

5

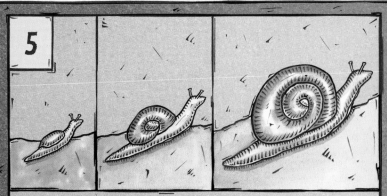

Each snail's shell grows with it, in a spiral shape. New shell material is added at the opening of the shell, and the part the snail was born with ends up right in the middle of the spiral.

BUSY BUZZERS

SOME BEES DON'T HAVE STINGERS—TRUE OR FALSE?

1 Bees don't buzz with their mouth! All the noise comes from their wings, whizzing up and down as fast as 200 times a second.

2 HONEYBEES are usually peaceful creatures, but they will sting to defend their nest — especially if anything tries to steal their honey!

3 A honeybee's stinger hooks in so tightly that it's left behind when the bee flies away. Parts of the bee's insides are torn away along with the stinger, and the poor bee dies.

4 Honeybees visit flowers to collect the two things they need to eat — sweet liquid nectar and powdery pollen.

8 BEES

5 The bees roll the pollen up into balls on their back legs — it looks as though they're wearing yellow pants.

6 A single nest can be home to 30,000 worker bees, who are all females. There's also one queen and a few hundred males, called drones.

7 Drones don't do much! Their only job is to mate with a queen when a new nest is started. Then they die.

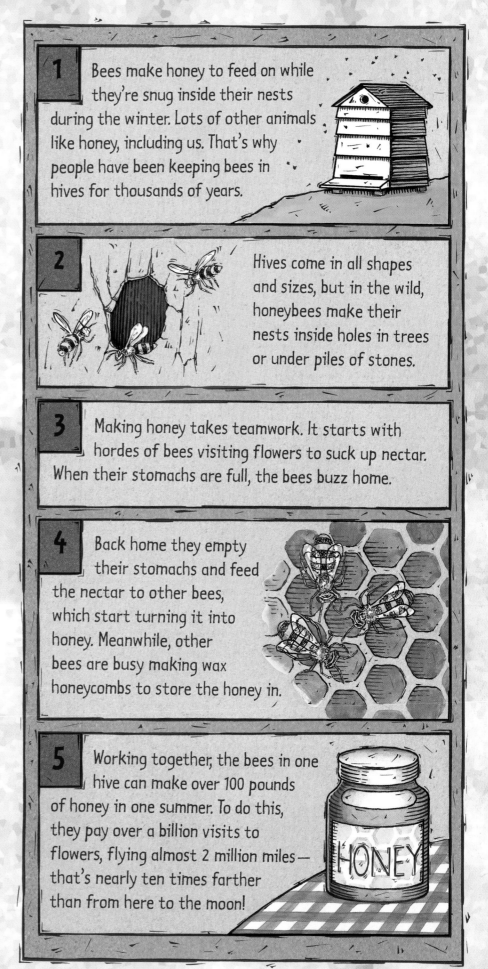

1 Bees make honey to feed on while they're snug inside their nests during the winter. Lots of other animals like honey, including us. That's why people have been keeping bees in hives for thousands of years.

2 Hives come in all shapes and sizes, but in the wild, honeybees make their nests inside holes in trees or under piles of stones.

3 Making honey takes teamwork. It starts with hordes of bees visiting flowers to suck up nectar. When their stomachs are full, the bees buzz home.

4 Back home they empty their stomachs and feed the nectar to other bees, which start turning it into honey. Meanwhile, other bees are busy making wax honeycombs to store the honey in.

5 Working together, the bees in one hive can make over 100 pounds of honey in one summer. To do this, they pay over a billion visits to flowers, flying almost 2 million miles — that's nearly ten times farther than from here to the moon!

1

Worms don't just live in the soil, they help to make it. At night, they crawl out of their tunnel looking for fallen leaves.

2 If a worm finds a leaf, it grips it in its mouth and drags it backward into its tunnel. When the worm is safely back underground, it starts to chomp through the soft parts of the leaf.

3

Worms have to go to the bathroom, just like you and me. They either leave their droppings underground or just outside their tunnel. You've probably seen these aboveground droppings — they're called wormcasts.

4 Wormcasts are full of nutrients that help to make the soil richer. And because there are so many worms and so many wormcasts, the soil is mostly made up of recycled worm droppings!

1 The soil is crawling with earthworms. There could be a million in an area the size of your school playground.

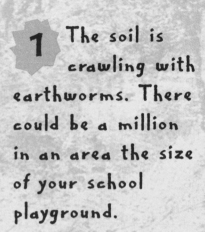

2 Although they feel rubbery, worms are actually covered with hundreds of tiny bristles. The bristles help them to grip as they wriggle along their tunnels.

WRIGGLY DIGGERS

5 The swollen part on an earthworm's body is called a saddle. It's closer to the head than to the tail, so you can use it to tell one end of a worm from the other.

3 Worms make tunnels by munching through the soil, swallowing tiny pieces of food along the way.

6 The world's longest earthworm comes from South Africa. It's not much thicker than your little finger — but at 23 feet long, it would outstretch three jump ropes!

4 If a bird tries to tug a worm out of the ground, the worm uses its bristles to grip on to its tunnel walls. Sometimes the bird wins the tug of war, sometimes the worm!

7 It's perfectly true that if you cut a worm in half, a new worm grows from each piece!

WORMS CAN TIE THEMSELVES IN KNOTS — TRUE OR FALSE?

BEASTLY BEAUTIES

1 Dragonflies are beautiful but deadly! They will hunt almost anything smaller than themselves— mosquitoes, flies, even bees and wasps.

2 Those great bulging eyes are all the better to see you with. Dragonflies have bigger eyes than any other insect. They probably have the best eyesight, too.

4 When they catch something, dragonflies wrap their bristly legs around it. Then they set to work with their powerful jaws, eating the unfortunate creature alive!

3 Dragonflies usually feed while flying, snatching their prey out of the air or up off the ground with their legs.

5 Some dragonflies can zip along at well over 20 miles per hour!

12 DRAGONFLIES

6 Most dragonflies lay their eggs in water. Some drop them out of the air like tiny bombs. Others just dip their tail into the water long enough to lay a single tiny egg.

You have only one lens in each eye to look through, but insects have lots and lots. Dragonflies have the most—as many as 30,000 per eye.

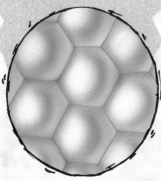

1 Dragonfly eggs hatch into nymphs. A nymph isn't at all like its beautiful parents—it lives underwater, has no wings, and is usually a boring brown.
The nymph spends its time lurking on the bottom of a river or pond, looking for tadpoles and other tasty food. If it spots something, it creeps forward, and then . . .

2 Zap! Out shoots its hooked lower lip. The tadpole is caught in a flash and dragged, still wriggling, toward the nymph's waiting jaws.

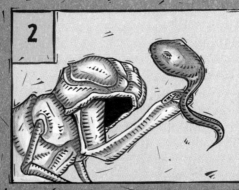

3 But nymphs are also hunted—by other underwater animals bigger than they are. If the nymph spots danger in time, it squirts water out of its bottom, shooting forward and dropping out of sight.

4 The nymph can spend years playing hide-and-seek like this. But one day, if it's lucky enough to survive, it crawls up a plant stem into the air. Its skin splits open, and out comes a glorious dragonfly!

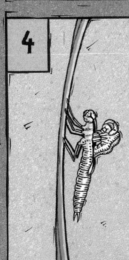

DRAGONFLIES 13

1 Not all spiders weave webs, but they all spin silk. Spiders make silk inside their body, pulling it out with their legs from holes near their bottom.

2 The silk is liquid when it comes out, but it instantly hardens into thread. A spider's silk is thinner than one of your hairs, but it's stronger than a steel wire of the same thickness.

3 Webs help spiders to catch their food. When an insect stumbles into a web, it soon becomes stuck. As it struggles to escape, it gets more and more tangled up.

4 When the spider feels its web trembling, it rushes out and bites the insect, poisoning it so that it can't move. Sometimes the spider wraps the insect in silk, and stores it alive in a corner of the web to eat later.

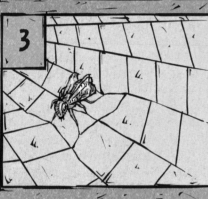

5 Webs soon get dirty and torn, so lots of spiders make a new one every day. They don't waste the old one, though—they roll it up into a ball and eat it!

14 SPIDERS

1 You'd get an awful shock if you found this furry monster in your bathtub! It's a BIRD-EATING SPIDER from Guyana in South America.

2 All bird-eating spiders are big, but this one's the world record holder. It's bigger than a dinner plate — more than 11 inches from leg tip to leg tip.

HAIRY AND SCARY

3 Bird-eating spiders don't usually catch their food in webs. Instead they prowl around the forest floor at night, hunting for their dinner.

5 When they find something tasty, bird-eating spiders grab it with their front legs. Then they bite it, injecting poison through their enormous fangs.

4 Despite their name, bird-eating spiders don't often eat birds. They're more likely to munch on lizards, beetles, and grasshoppers — or even other spiders!

6 Most spiders are very nearsighted. To make up for this, they use the hair on their body to feel their way around and to sense when other animals are near.

CHIRPY JUMPERS

1 There are over 20,000 different kinds of crickets and grasshoppers. Most of them chirp, and lots of them jump.

2 Only the males make any noise. They chirp to attract a female, and to tell other males not to get too close!

3 Some chirp by scraping their back legs along their front wings. Others rub their two front wings together.

This colorful grasshopper oozes a gross-tasting liquid if it's attacked. It's called the ELEGANT GRASSHOPPER and it lives in Africa.

16 GRASSHOPPERS

4 These insects' ears are in a very strange place — either on their stomach or their legs!

5 Those long back legs are very useful for escaping from enemies. Sometimes a cricket or grasshopper will leap into the air, spread its wings, and zoom off.

6 Usually, though, it will just jump a few feet away — leaving the hungry bird or lizard wondering why its lunch has vanished!

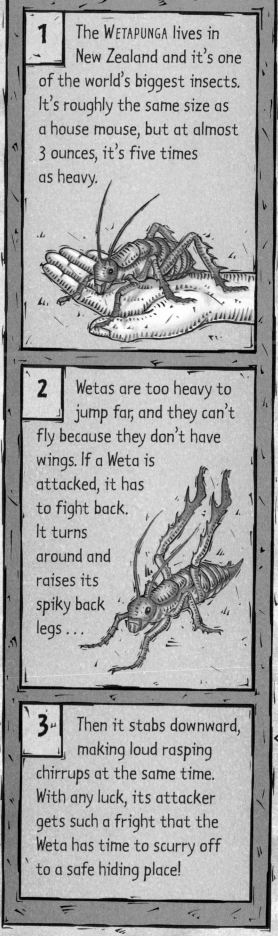

1 The WETAPUNGA lives in New Zealand and it's one of the world's biggest insects. It's roughly the same size as a house mouse, but at almost 3 ounces, it's five times as heavy.

2 Wetas are too heavy to jump far, and they can't fly because they don't have wings. If a Weta is attacked, it has to fight back. It turns around and raises its spiky back legs . . .

3 Then it stabs downward, making loud rasping chirrups at the same time. With any luck, its attacker gets such a fright that the Weta has time to scurry off to a safe hiding place!

AND CRICKETS 17

SOME ANTS DON'T MAKE NESTS—TRUE OR FALSE?

1 These busy rooms are in the middle of a WOOD ANTS' nest. The big ant is the queen, and all of the others are her daughters, the workers.

3 Worker ants are always scurrying around doing things. While some feed and clean the queen, others rush her eggs off to special nursery rooms.

2 The queen ant takes life easy. All she does is lay eggs — hour after hour, day after day. Her eggs hatch into ant grubs.

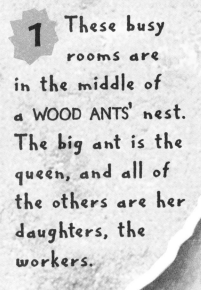

1 LEAFCUTTER ANTS live in North and South America, and their name tells you what they like doing best. At night they come out from their nest and swarm over trees, snipping out pieces of leaf with their scissorlike jaws.

2 Carrying the leaf pieces, the ants start back home. This can be quite a journey, often more than half a mile. If one ant gets tired and needs a rest, it passes its leaf to another one.

18 ANTS

WORKERS

4 Some of the worker ants also lay eggs. These aren't for hatching, though. They're fed to the queen and the ant grubs instead.

6 This is just part of the wood ants' home. The whole nest has hundreds of tiny rooms joined by tunnels.

5 Ants are very house-proud and work nonstop to keep their nest clean and tidy. Any trash is scooped up and dumped outside.

3 Back at the nest, the ants take their leaves deep underground to special rooms. Other ants now get to work—chopping the leaves up and rolling them into tiny balls.

4 A fuzzy white fungus soon grows all over the leaf balls. Leafcutter ants like this fungus so much that it's just about the only thing they eat. And with as many as a million ants in one nest, they go through a whole lot of it!

HARD CASES

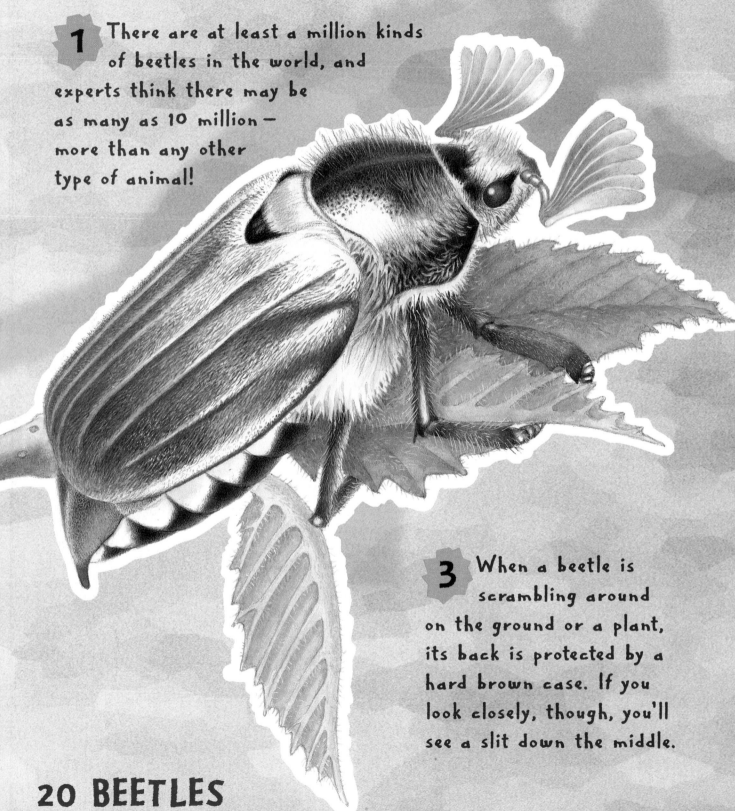

1 There are at least a million kinds of beetles in the world, and experts think there may be as many as 10 million — more than any other type of animal!

2 This is a European beetle called a COCKCHAFER. It lives in orchards, woods, and other places with lots of trees and shrubs.

3 When a beetle is scrambling around on the ground or a plant, its back is protected by a hard brown case. If you look closely, though, you'll see a slit down the middle.

1 At 7 inches long, male HERCULES BEETLES are the world's longest beetles. They look fierce, too—as if they have an enormous pair of jaws and might bite. In fact, these South American giants are harmless fruit eaters, and the only thing they attack is each other.

2 The jaws are actually a pair of horns. The bottom one grows out of the beetle's head, while the top one sprouts out of its back.

3 Male Hercules beetles fight to win burrows or patches of ground for mating. They circle around each other like wrestlers, trying to get a grip with their horns. Sometimes one of them manages to pick his opponent up and slam him down on his back.

4 The loser is left with his legs wriggling helplessly in the air. When he eventually gets back on his feet, he scuttles off to try to find another burrow. Meanwhile, the winner waits for any other Hercules beetle brave enough to take him on!

BEETLES 23

8 Most beetles are smaller than a baked bean — that's less than half an inch long. Cockchafers can grow to almost 2 inches, though. That's about as long as your thumb.

9 The world's heaviest beetles can weigh almost 4 ounces — nearly as much as a hamster. They're called GOLIATH BEETLES and they live in Africa.

BEETLES 22

4 Hiding under the case is a pair of delicate wings. When the beetle takes off, the two halves of the case flip forward. Then the beetle unfolds its wings and whirs away into the air.

5 The weird things sprouting out of the beetle's head are called antennae. All insects have antennae, and they use them for tasting, smelling, and feeling.

6 Antennae come in all shapes and sizes. There are short knobby ones, stiff bristly ones, and long trailing ones.

7 This LONGHORN BEETLE lives in Africa. Its antennae are at least twice as long as its body.

INDEX

Main illustrations by Sandra Doyle (cover, 3-5, 8-9, 12-13, 16-17);
Sarah Fox-Davies (20-22); Clive Pritchard (6-7, 10-11, 14-15, 18-19);
inset and picture-strip illustrations by Ian Thompson

Text copyright © 1996 by Martin Jenkins
Illustrations copyright © 1996 by Walker Books Ltd.

All rights reserved.

First U.S. edition 1996

Library of Congress Cataloging-in-Publication Data is available.
Library of Congress Catalog Card Number 95-53680

ISBN 0-7636-0036-9

2 4 6 8 10 9 7 5 3 1

Printed in Hong Kong

This book was typeset in Overweight Joe and Kosmik.

Candlewick Press
2067 Massachusetts Avenue
Cambridge, Massachusetts 02140

QUIZ ANSWERS

Page 2 — FALSE
A lot of male butterflies also drink salty things, like animals' sweat.

Page 6 — FALSE
Some snails, including little agate shells, give birth to live babies.

Page 8 — TRUE
Some kinds of bees don't have a stinger, and no male bees have a stinger.

Page 11 — TRUE
Worms tie themselves in knots if the soil gets very dry.

Page 12 — FALSE
The heaviest insect is the Goliath beetle ($4^{1}/_{2}$ oz.). The longest is the Indonesian giant stick insect (13 in.).

Page 15 — TRUE
That's how you can tell they're not insects, which have six legs. The two things next to the bird-eating spider's fangs are pedipalps— these are like insects' antennae.

Page 16 — TRUE
People in Asia keep pet crickets in special bamboo cages.

Page 18 — TRUE
Some ants, such as army ants from Africa and South America, live their whole lives on the move.

Page 23 — FALSE
Most beetles are vegetarians, but some kinds eat other animals.